MW00568114

This book belongs to

Ananya

ZIGZAG FACTFINDERS
COUNTRIES

Written by
Moira Butterfield

Consultant
Keith Lye

Edited by
Kay Barnham & Hazel Songhurst

zigzag

The consultant, Keith Lye, is a well-known geographer and author of over 90 books for children. He has also acted as consultant on many reference books and atlases.

ZIGZAG PUBLISHING

Published by Zigzag Publishing,
a division of Quadrillion Publishing Ltd.,
Godalming Business Centre, Woolsack Way,
Godalming, Surrey GU7 1XW, England.

Series concept: Tony Potter
Senior Editor: Nicola Wright
Design Manager: Kate Buxton
Designed by: Chrissie Sloan
Illustrated by: Shelagh McNicholas, Mainline Design,
Mike Lacey, Sally Reason, Pauline Jay,
Susan Robertson
Illustrations research: Paul Harrison
Production: Zoë Fawcett

Color separations: RCS Graphics Ltd, Leeds, England
Printed in Singapore

Distributed in the U.S. by SMITHMARK PUBLISHERS
a division of U.S. Media Holdings, Inc.,
16 East 32nd Street, New York, NY 10016

Copyright © 1997 Zigzag Publishing. First published in 1993.

All rights reserved. No part of this publication may be reproduced, stored in a retrieval system or transmitted by any means, electronic, mechanical, photocopying or otherwise, without the prior permission of the publisher.

ISBN 0-7651-9318-3
8023

Contents

About this book

This book takes you on a trip around the world! It is packed full of fascinating facts about different countries and illustrated with colorful maps and drawings. Find out the name of the highest waterfall in the world, where lions climb trees and where you can ride on a train made from Lego!

We have split the world into six areas. Each area is a different color. To find an area in the book, match the color at the top of the page.

This list explains some new or difficult words you will find in this book.

 A **border** divides one country from another.

 The **equator** is the imaginary line half-way around the Earth.

 Oil, gas and coal are **natural resources**.

 A **capital city** is the main city of a country.

 A **continent** is one of the world's main land masses.

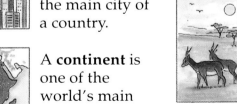 **Grasslands** are huge grassy plains, also called **prairies, pampas** and **savannah.**

 A **plateau** is high flat land.

 A **republic** has a president instead of a king or queen.

 A **delta** is where a river splits into streams before it meets the sea.

 Import and **export** is to carry goods **into** and **out of** a country.

 A **river basin** is the land drained by a river.

Europe

English — Hello!

French — Bonjour!

Europe is the second smallest continent in the world. About 700 million people live there, so it is very crowded compared to other regions.

Europe has ocean to the north, south and west. To the east, it borders the continent of Asia.

The Ural Mountains separate Europe from Asia.

There are more than 40 countries in Europe. The continent stretches from inside the Arctic Circle down to the Mediterranean Sea. Each country has its own government, capital city, languages and customs.

Key facts

Size: 4,062,158 sq. mi. (7% of the world's land surface)
Smallest country: Vatican City (.17 sq. mi.)
Largest country: Russia (1,757,141 sq. mi. of it is in Europe)
Longest river: Volga (2,194 mi.)
Highest mountain: Mount Elbrus (18,482 ft.) in the Caucasus mountain range

Countries

Twelve European countries are members of the EC, or the European Community. The members work together to make laws in areas such as farming, industry and finance.

The flag of the EC

The group of countries in the west of the European continent is sometimes known as Western Europe.

Many countries in the east of Europe are changing. The former Soviet Union has now divided into fifteen separate republics.

Landscape

Europe's landscape is very varied. In the far north you can see lots of forests and lakes. In central parts there are meadows and low hills. The south has some high mountain ranges and wide plains.

Western Europe's highest mountain range is the Alps, which stretches across the top of Italy. Alpine peaks are snowy all year. The Alps are very popular for skiing vacations.

Yassoo! Greek

Ola! Spanish

Halloj! Danish

Zdras-vuytya! Russian

Economy

There are lots of industries in Europe. Goods are imported and exported through the many seaports. Farming of all kinds is also important.

Europe has many busy seaports

Main industries

Fishing   Steel

Timber   Engineering

Farming   Mining

Textiles

Weather

In Scandinavia it is cold much of the time. In Eastern Europe the winters are very cold, but the summers are warm. Over Western Europe the summers are warm, the winters are cool and rain falls throughout the year.

Map key

1 Albania
2 Andorra
3 Austria
4 Belarus
5 Belgium
6 Bosnia & Herzegovina
7 Bulgaria
8 Croatia
9 Czech Republic
10 Denmark
11 Estonia
12 Finland
13 France
14 Germany
15 Gibraltar
16 Greece
17 Hungary
18 Iceland
19 Ireland
20 Italy
21 Latvia
22 Liechtenstein
23 Lithuania
24 Luxembourg
25 Macedonia
26 Malta
27 Moldova
28 Monaco
29 Netherlands
30 Norway
31 Poland
32 Portugal
33 Romania
34 Russia
(34) Kaliningrad
35 San Marino
36 Slovakia
37 Slovenia
38 Spain
39 Sweden
40 Switzerland
41 Ukraine
42 United Kingdom
43 Vatican City
44 Yugoslavia

Map of Europe

ARCTIC CIRCLE

SCANDINAVIA

Atlantic Ocean

Ural Mountains

North Sea

Baltic Sea

CENTRAL EUROPE

EASTERN EUROPE

WESTERN EUROPE

Alps

Mediterranean Sea

Black Sea

Caspian Sea

Caucasus Mountains

Approximate scale:
620 mi.
1,240 mi.

5

United Kingdom

The United Kingdom lies off Europe's north-west coast. It is made up of four countries.

Key facts

Size: 94,226 sq. mi.
Population: More than 57 million
Currency: Pound sterling
Main language: English
Also called: Britain, U.K.

England, Scotland, Northern Ireland and Wales make up the U.K.

English flag

Scottish flag

Welsh flag

Northern Irish flag

A London bus

Capital city: London. About 6.4 million people live here. London is the center of business and government.

Landscape: The highest mountains are in Wales and Scotland. The tallest is Ben Nevis (4,406 ft.) in Scotland. The longest river is the Severn (468 mi.). It flows from Wales into England.

Industries

Chemicals

Electronics

Textiles

Heavy machinery

Oil

Ben Nevis

Places to visit: There are lots of historical sites, ancient cities and towns. Britain is famous for its royal palaces and stately homes.

The Tower of London, Britain's most popular tourist attraction

France

France is one of the largest countries on the European continent.

Paris

Key facts

Size: 211,207 sq. mi.
Population: Over 56 million
Currency: French franc
Main language: French

Capital city: Paris. This is a world center of fashion and art. It has many famous art galleries and museums.

The Eiffel Tower is in Paris. It is made of iron and stands 984 ft. high. You can travel to the top by elevator.

Eiffel Tower

Landscape: France has many different kinds of scenery, with spectacular mountains, pretty river valleys and sunny beaches. Mont Blanc, on the Italian border, is the highest mountain (15,772 ft.). The longest river is the Loire (653 mi.).

Grapes are grown for wine

Industries

Farming

Wine

Tourism

Fashion

Vehicles

Chemicals

A vineyard

Places to visit: You can ski in the Alps, swim in the Mediterranean or visit many historic chateaux. EuroDisney™ is near Paris.

The royal château at Versailles

Spain

Spain is in south-west Europe. It is the third largest country on the continent.

Key facts

Size: 194,896 sq. mi.
Population: Over 38 million
Currency: Peseta
Main language: Spanish

The Royal Palace

Capital city: Madrid. This is a famous center of culture with many theaters, cinemas and opera houses.

Landscape: Spain is a mountainous country. In the center there is a vast, high plateau. The highest mountain is Mt. Mulhacen (11,440 ft.). The longest river is the Tagus (626 mi.).

The Sierra Nevada mountains

Places to visit: There are many historic cities and palaces built by the Moors, who invaded from Africa in 711. In the south there are sunny beaches.

Industries

Tourism
Wine
Farming
Vehicles
Chemicals
Electronics

The Alhambra, a Moorish palace near Granada

Germany

Germany borders nine other countries. In 1990, East and West Germany joined to become one country.

Key facts

Size: 137,743 sq. mi.
Population: About 79 million
Currency: Deutsche Mark
Main language: German
Full name: Federal Republic of Germany

Capital city: Berlin. This city was once divided by a high wall mounted with guns. Today, people can go wherever they like in the city.

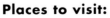

The Brandenburg Gate, Berlin

Black Forest pinewoods

Landscape: There are many different kinds of scenery. The beautiful Rhine River Valley and the Black Forest are very famous. The Zugspitze (9,722 ft.) is the highest mountain. The longest river is the Elbe (724 mi.).

Industries

Chemicals
Vehicles
Engineering
Coal
Shipbuilding

Neuschwanstein Castle, Bavaria

Places to visit: There are historic cities and ancient castles in the regions of Bavaria and Saxony. In the north there are sandy beaches.

Norway

Norway lies along the coast of Scandinavia. Part of it is inside the Arctic Circle.

● Oslo

Key facts

Size: 125,181 sq. mi.
Population: Over 4 million
Currency: Norwegian krone
Language: Norwegian

Capital city: Oslo. This is one of the world's largest cities, but only 500,000 people live here. You can visit many ancient Viking burial mounds and settlement sites in Norway.

Landscape: Norway has a long coastline, famous for its deep inlets called fjords. It also has over 150,000 islands. Mountains and moorland cover three-quarters of the country. Glittertind is the highest mountain (8,104 ft.). The longest river is the Glama (373 mi.).

A Norwegian fjord

Places to visit: The Norwegian mountains are famous for winter sports. There are about 10,000 ski jumps in the country, as well as lots of forest trails.

Industries

Oil
Paper-making
Timber
Fishing

Cross-country skiing in Norway

Sweden

Sweden is the largest Scandinavian country. It has a long coastline and many islands.

● Stockholm

Key facts

Size: 173,731 sq. mi.
Population: Over 8 million
Currency: Swedish krona
Main language: Swedish

Capital city: Stockholm. The city is built on a string of islands. It is the home of the Royal Palace and many other historic buildings.

The Royal Palace

Landscape: Over half of Sweden is covered with forest. There are about 96,000 lakes in the south and centre of the country. Lapland (northern Sweden) is inside the Arctic Circle. Mount Kebnerkaise (6,926 ft.) is the highest peak in Sweden.

Places to visit: Sweden has thousands of islands which are ideal for boating and fishing. You can find out about Viking longboats in the Nordic Museum, Stockholm.

Industries

Timber
Vehicles
Electronics
Minerals
Chemicals

A Viking longboat

Denmark

Denmark is the smallest Scandinavian country. It is made up of a peninsula and about 400 islands.

Copenhagen

Key facts

Size: 16,629 sq. mi.
Population: Over 5 million
Currency: Danish krone
Main Language: Danish

Capital city: Copenhagen. This is a city with many old buildings, fountains and pretty squares. The well-known statue of the Little Mermaid sits on a rock in Copenhagen Harbor. The story of the mermaid was written by a famous Dane, Hans Christian Andersen.

The Little Mermaid

Landscape: Denmark has mainly low-lying countryside with forests and lakes. There are many beautiful sandy beaches and about 500 islands. Yding Skovhoj is the highest mountain (568 ft.). The longest river is the Guden (98 mi.).

Places to visit: Copenhagen has palaces, castles and a Viking museum. At Legoland™ Park everything is made of Lego™, including life-sized working trains and lots of miniature buildings.

Industries

Farming
Tourism
Textiles
Electronics
Oil

A Legoland™ train

The Netherlands

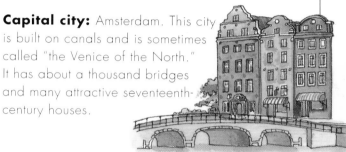

The Netherlands is one of the flattest European countries. Two-fifths is below sea level.

Amsterdam

Key facts

Size: 15,770 sq. mi.
Population: Over 15 million
Currency: Guilder
Main language: Dutch
Also called: Holland

Capital city: Amsterdam. This city is built on canals and is sometimes called "the Venice of the North." It has about a thousand bridges and many attractive seventeenth-century houses.

Amsterdam

Landscape: The Netherlands is very flat and is criss-crossed by rivers and canals. In the past, large areas of land were reclaimed from the sea. Sea-dams, called dykes, were built and the sea water was drained away. The highest point of land is only 1,056 ft.

Places to visit: The tulip fields are a world-famous sight. In the countryside there are lots of pretty old towns and villages, country houses and castles. Windmills are still used in parts of the country.

Industries

Flowers
Farming
Diamond-cutting
Electronics
Chemicals

Tulips are exported to many parts of the world

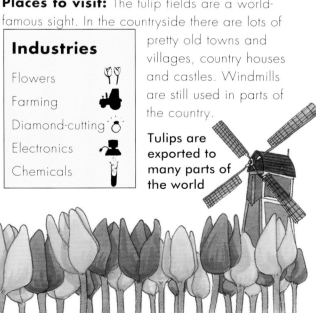

Italy

Italy is in southern Europe. The Mediterranean islands of Sicily and Sardinia are part of this country.

Key facts

Size: 116,303 sq. mi.
Population: Over 57 million
Currency: Lira
Main language: Italian

Capital city: Rome. This was once the capital of the ancient Roman empire. It has lots of Roman remains, including the ruins of the Colosseum. Here, huge audiences watched as gladiators fought and Christians were thrown to the lions.

The Colosseum

In Rome there is a tiny separate country called Vatican City. This is the home of the Pope, the head of the Roman Catholic Church.

Landscape: There are spectacular mountains and beautiful lakes in the north. In central and southern regions you can see plains and smaller mountains. Italy has several live volcanoes, including Etna, Vesuvius and Stromboli. Mont Blanc is the highest mountain (15,771 ft.). The Po is the longest river (405 mi.).

Stromboli

Industries

Farming	
Vehicles	
Electronics	
Fashion	

Places to visit: Italy has many historic cities, such as Venice, which is built on a lagoon. The easiest way to travel around Venice is by gondola. Some of the world's greatest artists have lived in Italy. You can see their work in art galleries and museums.

A Venetian gondola

Greece

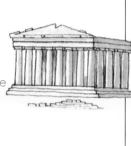

Greece juts out into the Mediterranean Sea. About one-fifth of the country is made up of small islands.

Key facts

Size: 50,949 sq. mi.
Population: Over 10 million
Currency: Drachma
Main language: Greek

Capital city: Athens. This is one of the world's oldest cities. On the top of the Acropolis ridge, you can see the ruins of the Parthenon. This Greek temple is 2,400 years old.

Landscape: Mainland Greece has plains and forests in the south and is mountainous in the north. Mount Olympus is the highest mountain (9,571 ft.). The Greek islands vary in size and landscape. Crete is the largest of these. The longest river in Greece is the Aliakmon (184 mi.).

Mount Olympus, legendary home of the ancient Greek gods

Industries

Tourism	
Fishing	
Farming	

Places to visit: There are lots of ancient sites. Many of them are linked with stories from Greek mythology. For instance, the Palace of Knossos, on Crete, is the legendary home of the Minotaur monster.

The Minotaur

Poland

Poland is in Central Europe. It shares its borders with four other countries. Its coastline is on the Baltic Sea.

Key facts

Size: 120,725 sq. mi.
Population: Over 38 million
Currency: Zloty
Main language: Polish

Capital city: Warsaw. The old city of Warsaw was destroyed in World War Two. The "Old Town" area has been rebult in the style of the old buildings.

Warsaw "Old Town"

Landscape: There are many lakes and wooded hills in northern Poland, and beach resorts along the Baltic coast. Rysy Peak (8,212 ft.) is the highest point in the mountainous south. The longest river is the Vistula (664 mi.).

Industries

Farming
Coalmining
Shipbuilding
Timber

Places to visit: There are lots of museums and art galleries in Warsaw. There are several national parks where rare forest creatures live, such as lynxes and moose.

A rare lynx

Russia

Russia is the largest country in the world. About a quarter lies in Europe. The rest is in Asia.

Key facts

Size: 6,592,812 sq. mi.
Population: About 150 million
Currency: Rouble
Main language: Russian

Capital city: Moscow. The famous Kremlin building is in the center of the city. It was once a fortress occupied by Tsars, Russian emperors, who ruled for centuries.

St. Basil's Cathedral, Moscow's famous landmark

Landscape: Russia is the world's largest country. It has some of the largest lakes and forests and longest rivers in the world. Mount Elbrus is the highest mountain (18,482 ft.). The longest river is the Volga (2,193 mi.).

Industries

Engineering
Farming
Oil
Minerals
Coal mining

Lake Baikal, is the world's deepest lake, at 6,365 ft.

Places to visit: Russia has many old cities, all with long and exciting histories. The world's longest railway, the Trans-Siberian, runs across the country from Moscow to Vladivostok.

The Trans-Siberian Railway

11

North America

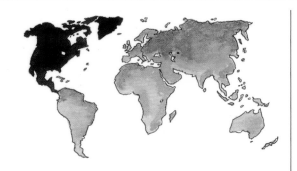

North America is the third largest continent in the world. It stretches from the frozen Arctic Circle down to the sunny Gulf of Mexico. It is so wide that there are eight different time zones.

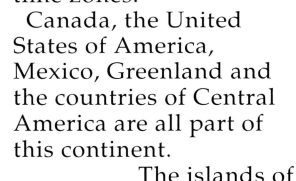

The frozen north

Canada, the United States of America, Mexico, Greenland and the countries of Central America are all part of this continent.

The islands of Hawaii, Bermuda and the West Indies lie off the mainland.

The tropical south

Key facts

Size: 9,035,000 sq. mi.
Largest country: Canada (second largest in the world) 3,851,787 sq. mi.
Longest river: Mississippi-Missouri (3,860 mi.)
Highest mountain: Mount McKinley (20,322 ft.)

Landscape

The North American landscape is very varied. There are huge forests, ice-covered wastelands, scorching deserts and wide grassy plains.

The Rocky Mountains

In the west, mountains run from Alaska down to Mexico. They are called the Cordillera, and include the Rocky Mountains. The Appalachian Mountains run down the eastern side.

A vast plain stretches about 2,983 mi. from the Gulf of Mexico to northern Canada. The Great Plains and the Mississippi-Missouri River are in this region. Much of the world's maize is grown here.

The most northern parts of the continent border the icy Arctic Ocean. This area is called tundra. Here, the land just beneath the surface is permanently frozen.

Weather

Temperatures vary from -164°F in an Arctic winter to 137°F in summer in Death Valley, California, one of the world's hottest places. Every year in North America, there are about 550 tornadoes, mostly in the central states of the U.S. Tornado winds can spin at up to 403 mph.

A tornado

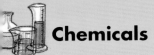 
People

Long ago, North America was populated only by native American people, called Native American Indiains. Over the centuries, settlers came from Europe and slaves were brought from Africa. Now, most of the population is descended from these people.

For many centuries, the Canadian Arctic has been the home of the Inuit people.

Many people from the West Indies are descended from Africans captured and brought over to the country as slaves.

The Statue of Liberty in New York City Harbor was the first sight many immigrants had of North America

Economy

The continent of North America has many natural resources, such as oil, minerals and timber.

The U.S. is the world's richest country. It has many different industries.

Timber products, such as paper, come from the northern areas

Cities

The U.S. has many cities. New York City is the largest, with a population of over 17 million.

Although Canada is the biggest country in North America, it has a small population. Its major cities are all in the south of the country, where the climate is milder than the Arctic north.

Mexico City, in Mexico, is one of the most overcrowded cities in the world.

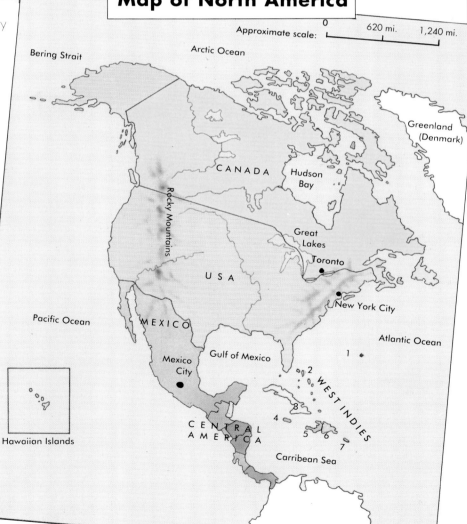

Map of North America

Approximate scale: 0 620 mi. 1,240 mi.

Bering Strait
Arctic Ocean
Greenland (Denmark)
CANADA
Hudson Bay
Rocky Mountains
Great Lakes
Toronto
USA
New York City
Atlantic Ocean
Pacific Ocean
MEXICO
Mexico City
Gulf of Mexico
WEST INDIES
CENTRAL AMERICA
Caribbean Sea
Hawaiian Islands

Map key

1 Bermuda
2 Bahamas
3 Cuba
4 Jamaica
5 Haiti
6 Dominican Republic
7 Puerto Rico

U.S.A.

The United States of America stretches from the Atlantic Ocean to the Pacific. It includes Alaska in the north and the Hawaiian Islands in the Pacific Ocean.

Washington D.C.

Key facts

Size: 3,618,766 sq. mi.
Population: Over 250 million
Currency: U.S. dollar
Main language: English

Capital city: Washington, D.C. This is the center of government. The President lives here in the White House.

The White House

The U.S. flag is called the "Stars and Stripes". It has thirteen red and white stripes which stand for the first thirteen states. The country now has fifty states, shown by the fifty stars on the flag.

Landscape: As well as mountain ranges, vast plains and deserts, there is the deep land gorge, the Grand Canyon, in Arizona. Mount McKinley is the highest point (20,322 ft.). The Mississippi-Missouri (3,860 mi.) is the longest river.

Grand Canyon

Industries

Farming
Oil
Steel
Vehicles
Space

Places to visit: The U.S. has many historic sites, big cities and beach resorts. Famous places include Hollywood in California, Disney World™ in Florida, and many national parks.

The famous Hollywood sign

Canada

Canada is the largest country in North America. Its northern part is inside the Arctic Circle.

Ottowa

Key facts

Size: 3,851,787 sq. mi.
Population: About 27 million
Currency: Canadian dollar
Main languages: English and French

Capital city: Ottowa. Canada is divided into ten provinces and two territories. Ottowa is in the province of Ontario.

CN Tower

The CN Tower, the world's tallest free-standing structure (1,814 ft. high), is in Toronto, the largest city in Canada.

Landscape: Nearly half of Canada is covered by forest. The Great Lakes, the world's largest group of freshwater lakes, are on the border with the U.S. The Rocky Mountains are in the west. In the center, there are vast plains, called prairies. The highest mountain is Mount Logan (19,525 ft.). The Mackenzie is the longest river (2,635 mi.).

The Rocky Mountains

A totem pole

Industries

Farming
Forestry
Vehicles

Places to visit: You can see ancient totem poles in Stanley Park, Vancouver. Canada has lots of national park areas, where bears and wolves live. The world-famous Niagara Falls is a spectacular sight.

Mexico

Mexico lies south of the U.S.A. and north of South America.

Mexico City

Key facts

Size: 692,102 sq. mi.
Population: Over 80 million
Currency: Peso
Main language: Spanish

Capital city: Mexico City. Almost one-fifth of the population lives here. There are modern buildings next to ancient Aztec ruins.

Mexico once belonged to Spain. You can still see many Spanish-style buildings and churches.

A Spanish-style church

Landscape: More than half the country is over 3,200 ft. high. Central Mexico is a plateau surrounded by volcanic mountains. There are also deserts and swamps. Mount Orizaba (18,702 ft.) is the highest mountain. The longest river is the Rio Grande (1,299 mi.), which flows along the U.S. border.

Industries

Coffee
Oil
Minerals
Crafts

Places to visit: Aztec, Mayan and Toltec people once ruled Mexico. You can still see the remains of cities and temples they built.

A Mayan temple

Jamaica

Jamaica is an island in the Caribbean Sea, south of Cuba.

Kingston

Key facts

Size: 4,244 sq. mi.
Population: Over 2 million
Currency: Jamaican dollar
Main language: English

Capital city: Kingston. Built by a deep, sheltered harbor, this city was once ruled by a pirate, Captain Morgan, and his Buccaneers.

Landscape: Jamaica is a tropical island with lush rainforests, pretty waterfalls and dazzling white beaches. It is actually the tip of an undersea mountain range. Blue Mountain Peak (7,401 ft.) is the highest mountain on the island.

Captain Morgan

Places to visit: As well as beaches, there are wildlife parks and bird sanctuaries where some of the world's most exotic birds can be seen. There are working sugar and banana plantations.

Industries

Tourism
Minerals
Farming

A Jamaican beach

15

South America

South America is the world's fourth largest continent. It stretches from the border of Central America to the tip of Chile.

There are 13 South American countries, each with its own distinctive type of landscape and culture.

For centuries the continent was populated only by native peoples. Europeans did not arrive until 1499. Now, many of the people are descended from these Spanish or Portuguese settlers.

Early settlers came from Europe

A Spanish-style church

Landscape

There are lots of different landscapes, including high mountains, hot and cold deserts, rainforests and plains.

The world's longest mountain range is the Andes, which runs all the way down the western side. Some of its mountains are live volcanoes. There are also regular earthquakes in this area.

Between the highland areas there are vast lowlands. Much of these areas are covered by dense rainforests. There are also grasslands, such as the Argentinian pampas.

There are some live volcanoes in the Andes Mountain Range

South America has many dense rainforests

People

Many large South American cities are on the eastern coast. Outside the cities, most people live on small farms, growing just enough food to feed themselves.

There are still a few Amerindians (native Americans) living in forest settlements deep in the Amazon River Basin. They hunt animals, gather fruit and grow crops to eat.

Largest country: Brazil (3,286,470 sq. mi.)

Smallest country: French Guiana (38,749 sq. mi.)

Highest mountain: Mount Aconcagua (22,836 ft.)

Longest river: Amazon (4,000 mi.)

Amazon butterflies

A jaguar

The Amazon

The Amazon River carries more fresh water than any other. It flows across South America from the Andes to the Atlantic and drains more than 2 million sq. mi. of land, much of it rainforest. It is one of the world's richest wildlife areas, with many extraordinary creatures such as hummingbirds, sloths, jaguars and piranha fish.

A hummingbird

Weather

In rainforest areas, it is warm and humid all the time and it rains almost daily.

The warmest part of South America is in northern Argentina. The coldest place is Tierra Del Fuego, which faces Antarctica at the southern tip of the continent.

The world's driest place is the Atacama Desert in Chile. Until 1971 it had not rained there for 400 years.

Industries

Coffee
Forestry
Cacao
Oil
Farming
Minerals

Map of South America

Approximate scale:
0   620 mi.   1,240 mi.

Caracas
VENEZUELA
Orinoco River
COLOMBIA
2
3
4
1
PERU
Amazon River
EQUATOR
BRAZIL
Lima
La Paz
Brasilia
BOLIVIA
Pacific Ocean
CHILE
6
São Paulo
Paraná River
Santiago
5
Buenos Aires
ARGENTINA
Atlantic Ocean
Falkland Islands (U.K.)
Tierra del Fuego
(Chile) (Argentina)

Map key

1 Equador
2 Guyana
3 Surinam
4 French Guiana
5 Uruguay
6 Paraguay

Brazil

Brazil covers almost half of South America. It is the fifth-largest country in the world.

Brasilia

Key facts

Size: 3,286,470 sq. mi.
Population: Over 150 million
Currency: Cruzeiro
Main language: Portuguese

Capital city: Brasilia. This city was begun in the 1950s. It is famous for its futuristic architecture.

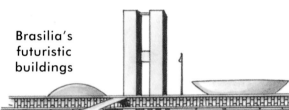

Brasilia's futuristic buildings

Landscape: Brazil has over 1 million sq. mi. of rainforest. The mighty Amazon River (4,000 mi.) runs through it, carrying more water than any other river in the world. Pica da Neblina (9,889 ft.) is the highest mountain in Brazil.

Flesh-eating piranha fish live in the Amazon River

Industries

Coffee
Sugar cane
Timber
Iron
Precious gems

Places to visit: The Amazon Rainforest has many thousands of animal and plant species. It is also the home of the Amazon Indians. Ecologists are trying to save the forest from destruction.

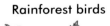

Rainforest birds

Argentina

Argentina lies to the east of the Andes Mountains, facing the Atlantic Ocean.

Buenos Aires

Key facts

Size: 1,068,296 sq. mi.
Population: Over 32 million
Currency: Austral
Main language: Spanish

Capital city: Buenos Aires. This is one of the largest cities in the southern half of the world. It is the birthplace of the famous tango dance.

Dancing the tango

Landscape: The scenery includes hot desert, the Andes Mountain Range and the cold wilderness of the Patagonia Desert in the south. Mount Aconcagua (22,836 ft.) is the highest point. The longest river is the Paraná (3,032 mi.).

Argentinian grassland, called pampas

Places to visit: There are historic cities, beach resorts and wildlife parks in Argentina. The Andes is home to the spectacular and rare bird, the condor.

Industries

Farming
Textiles
Steel
Chemicals

A condor

Peru

Peru is on the Pacific coast of South America. The Andes Mountains run down the center.

Lima

Key facts

Size: 496,222 sq. mi.
Population: Over 22 million
Currency: New Sol
Main languages: Spanish, Quechua

Capital city: Lima. This city was founded by the Spaniard, Francisco Pizarro, in 1535. He attacked and conquered Peru in the 1500s in search of its legendary treasure.

Francisco Pizarro

Landscape: Peru is a country of deserts, mountains and rainforests, some of it still unexplored. The source of the Amazon River is in the Peruvian Andes. The highest point is Mount Huascarán (22,206 ft.). The Ucayli is the longest river (910 mi.).

Industries

Farming
Coffee
Cotton
Fishing
Minerals

The Andes Mountains

Places to visit: Peru was once ruled by the Inca people. They built fabulous palaces, towers and temples covered in gold. You can visit the ruins of their cities, such as Cuzco and Machu Picchu.

Machu Picchu

Venezuela

Venezuela is on the north coast of South America. This is where Christopher Columbus first set foot on the American mainland.

Caracas

Key facts

Size: 352,142 sq. mi.
Population: Over 20 million
Currency: Bolívar
Main language: Spanish

Capital city: Caracas. The national hero, Simon Bolívar, is buried here. He helped to free Venezuela from the Spanish. Many buildings and streets are named after him.

Simon Bolívar

Landscape: There are lowland and highland areas. One of the lowland areas is a dense alligator-infested forest around the Orinoco River. Some of the highlands are still unexplored. Mount Bolívar is the highest point (16,428 ft.). The Orinoco (1,700 mi.) is the longest river.

Industries

Fruit
Coffee
Minerals
Tourism

The Venezuelan landscape

Places to visit: Venezuela has some of the most spectacular scenery in the world. the world's highest waterfall, Angel Falls (3,212 ft.) is in the Canaima National Park.

Angel Falls

Africa

Africa is the second largest continent in the world. It is only 8 mi. to the south of Europe, across the Strait of Gibraltar. A strip of land called the Isthmus of Suez separates the continent from Asia.

There are 670 million people in Africa, but they are spread thinly throughout more than 50 countries. Large parts of Africa are uninhabited because the climate is harsh and the terrain makes travel difficult. There are mountains, rainforests, deserts and grassland, called savannah.

Nairobi, Kenya

African desert

Key facts

Size: 11,684,158 sq. mi.
Population: 670 million
Largest country: Sudan
Smallest country: Seychelles
Highest mountain: Mount Kilimanjaro (19,341 ft.)
Longest river: Nile (4,145 mi.)

Mount Kilimanjaro

Landscape

Much of Africa is on a high plateau. Around this high, flat tableland there are narrow coastal plains. The highest mountains are in East Africa.

The Sahara Desert crosses the northern part of Africa. It is the biggest desert in the world, spreading out over a huge area almost as large as the U.S.A! Some parts of it are sandy, but much of it is rocky wasteland.

In central Africa, there are dense tropical rainforests. In southern Africa there are savannah and desert areas.

The Nile River is the longest river in the world. It flows north to the Mediterranean Sea.

The Nile River

Animals

Africa is very rich in wildlife. Many of the large mammals live on the wide grassy plains of the savannah.

Some species of animals are in danger of dying out. To save them, large areas have been made into reserves and national parks, where hunting is illegal.

African savannah

20

Weather

The Equator crosses Africa. Here there are hot, humid rainforests where it rains almost every day.

African rainforest

People

In the north, many people speak Arabic and follow the Muslim religion.

Pygmy people live in the rainforests of Congo and Zaïre.

In South Africa, black people make up two-thirds of the population. For many years they had few rights. This is now beginning to change.

Economy

Many Africans are farmers. They work on the land growing crops such as peanuts and cocoa beans (used for making chocolate).

A major industry is mining. Africa is rich in minerals such as gold, silver, tin and copper. It has large stores of oil and natural gas.

A farmer harvesting his crop

Map of Africa

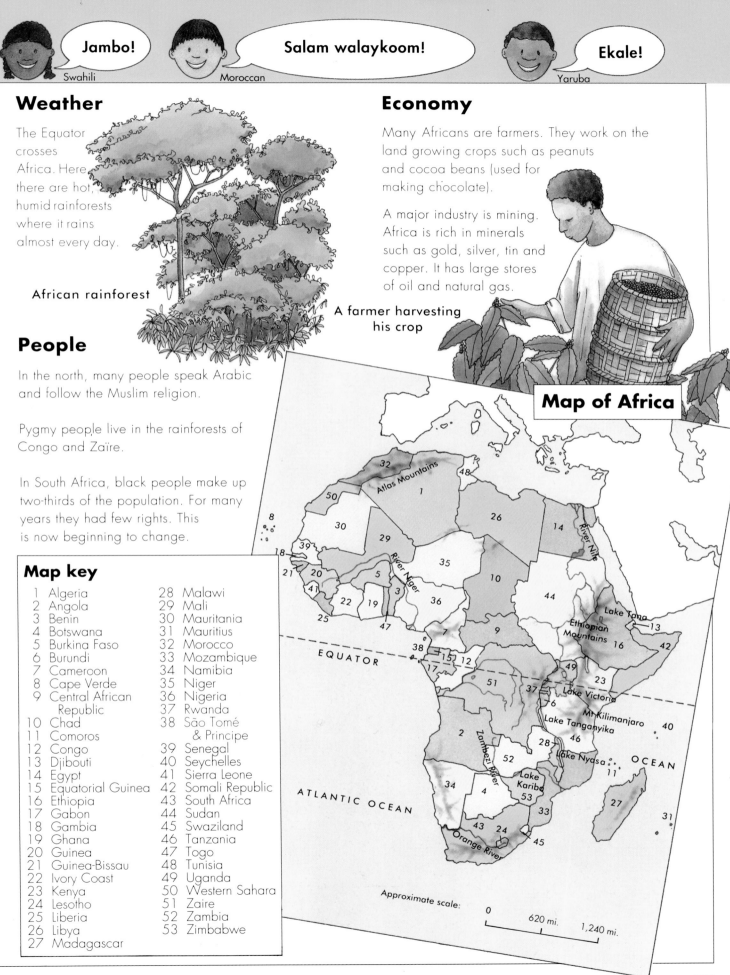

Map key

1 Algeria
2 Angola
3 Benin
4 Botswana
5 Burkina Faso
6 Burundi
7 Cameroon
8 Cape Verde
9 Central African Republic
10 Chad
11 Comoros
12 Congo
13 Djibouti
14 Egypt
15 Equatorial Guinea
16 Ethiopia
17 Gabon
18 Gambia
19 Ghana
20 Guinea
21 Guinea-Bissau
22 Ivory Coast
23 Kenya
24 Lesotho
25 Liberia
26 Libya
27 Madagascar
28 Malawi
29 Mali
30 Mauritania
31 Mauritius
32 Morocco
33 Mozambique
34 Namibia
35 Niger
36 Nigeria
37 Rwanda
38 São Tomé & Principe
39 Senegal
40 Seychelles
41 Sierra Leone
42 Somali Republic
43 South Africa
44 Sudan
45 Swaziland
46 Tanzania
47 Togo
48 Tunisia
49 Uganda
50 Western Sahara
51 Zaire
52 Zambia
53 Zimbabwe

Approximate scale:

0   620 mi.   1,240 mi.

21

South Africa

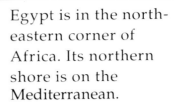

South Africa is at the southernmost tip of the African continent.

Key facts
Size: 471,442 sq. mi.
Population: Over 40 million
Currency: Rand
Main languages: English, Afrikaans.

Capital cities: South Africa has three capital cities called Pretoria, Bloemfontein and Cape Town. Each plays a different part in the government of the country.

Table Mountain, near Cape Town

Landscape: South Africa has a huge high plateau bordered by mountains. On the plateau there are wide grassy plains called the "veld." The long coastline has lots of beautiful beaches. The highest point is Champagne Castle in the Drakensberg (11,074 ft.). The Orange River (1,305 mi.) is the longest river.

Industries
Mining
Precious gems
Farming

South Africa is famous for its diamond mines

Places to visit: There are several famous wildlife reserves. You can see eagles, elephants, giraffes, lions and tigers living in their natural habitat.

A giraffe

Egypt

Egypt is in the north-eastern corner of Africa. Its northern shore is on the Mediterranean.

Key facts
Size: 386,659 sq. mi.
Population: Over 55 million
Currency: Egyptian pound
Main language: Arabic

Capital city: Cairo. This is a center of historic monuments and unique historical sites.

The Ancient Egyptian civilization began about 5,000 years ago. The Pyramids at Giza were built as tombs for the Pharoahs, the kings who ruled at that time.

The pyramids and Sphinx at Giza, just outside Cairo

Landscape:
The country is divided by the great Nile River. A belt of green fertile land runs along the river and spreads out around its delta (river mouth). The rest of Egypt is sandy desert. The highest mountain is Jabal Katrinah (8,652 ft.). The Nile River is the world's longest river (4,000 mi.).

Farming the Nile delta

Industries
Tourism
Textiles
Oil

Places to visit: Some pharoahs were buried in the Valley of the Kings. You can see their tomb treasures on display. There are huge temples at Luxor and Abu Simbel. Around the Red Sea there are beaches and spectacular coral reefs.

The great temple of Abu Simbel

Nigeria

Nigeria's coast is on the Gulf of Guinea in western Africa.

Abuja

Key facts

Size: 356,667 sq. mi.
Population: Over 88 million
Currency: Naira
Main language: The official language is English but there are over 250 local languages

Capital city: In 1991 Abuja became the new capital city, in place of Lagos.

Many people live in villages. The main groups of people are the Hausa, Ibo and Yoruba.

A village market

Landscape: There are lagoons, beaches and mangrove swamps along the coast. Inland, there are rainforests where the trees can grow as high as 100 ft. An area of grassland with scattered trees, called the savannah, lies in the south.

Coastal mangrove swamp

Industries
Oil
Cacao
Palm oil
Minerals

Places to visit: Many ancient cultures flourished in Nigeria. There are several historic cities, such as Kano, famous for its festival of horsemanship. There are wildlife parks and spectacular scenery.

A Yoruba festival mask and drummer

Tanzania

Tanzania lies in eastern Africa on the Indian Ocean. It includes the islands of Zanzibar and Pemba.

Dodoma

Key facts

Size: 364,896 sq. mi.
Population: About 25 million
Currency: Tanzanian shilling
Main languages: English, Swahili

Capital city: Dodoma. Dar es Salaam is the largest city and main port.

There are about 120 groups of people in Tanzania including the wandering Masai, the tallest people in the world.

Masai women

Landscape: Most of Tanzania consists of a high plateau. Savannah and bush cover about half the country. There are huge inland lakes. Mount Kilimanjaro (19,341 ft.), in Nigeria, is the highest mountain in Africa. The longest river is the Rufiji.

An inland lake

Industries
Farming
Cotton
Coffee
Minerals

Places to visit: Some of the world's finest wildlife parks are here, including the Serengeti Plain and the Ngorongoro volcano.

Lake Manyara is famous for its tree-climbing lions

Western & Southern Asia

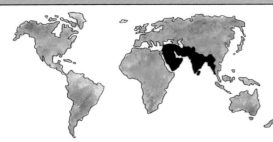

This area stretches about 3,107 mi. from the Mediterranean Sea to the Bay of Bengal. It includes Middle Eastern countries such as Saudi Arabia and the south Asian countries of Pakistan, Afghanistan, India and Bangladesh. The island of Sri Lanka and the mountainous countries of Bhutan and Nepal are also in this vast area.

Middle Eastern desert

The landscape varies from barren desert to the Himalayas, the highest mountains in the world.

The mountains of Nepal

Key facts

Size: 4,420,845 sq. mi.
Population: About 1330 million
Longest river: Ganges (2,900 mi.)
Highest mountain: Mount Everest (29,030 ft.)
Largest country: India (1,269,338 sq. mi.)
Smallest country: Maldives (115 sq. mi.)

Landscape

The landscape of Western and Southern Asia includes high mountains, wide plains, deserts and rainforests.

Mountains run all the way from Turkey to Afghanistan and dry, scrubby grassland stretches from Pakistan to Syria.

Desert covers most of Saudi Arabia and the surrounding lands.

Weather

The mountainous northern areas of the Middle East have hot summers and freezing winters. Farther south, it is hot all year round.

In India and Bangladesh, the farmers rely on the monsoon for their crops to grow. The monsoon is a season of heavy rain that falls between June and October.

Middle Eastern desert and mountains in Afghanistan

Monsoon rainfall in Southern Asia

Marhabah assalamu aleikum! — Arabic

Namaste! — Hindi

Min ga la baa! — Burmese

People

There are many different cultures in this area. Here are some examples:

In the Middle East, many people follow the Muslim faith. They pray every day facing the Holy City of Mecca, the center of their religion.

Many Indians are Hindus. They worship several gods, the chief of which is called Brahman. Some Indians follow the Sikh faith. Many Sikhs live in the Punjab region.

To Hindus, cows are sacred animals and are allowed to graze wherever they like

Economy

The countries around the Persian Gulf, in the Middle East, have vast oil reserves. They depend on the money they make from selling this oil around the world.

Throughout Southern Asia many different crops are grown, including wheat, millet, rice and cotton. India and Sri Lanka are among the world's biggest tea producers.

Tea picking

Main industries

Oil		Farming		
Mining		Textiles		

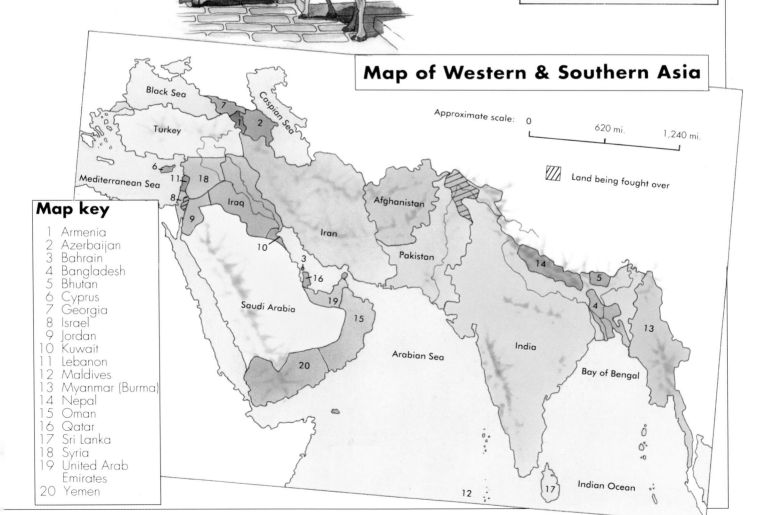

Map of Western & Southern Asia

Approximate scale: 0 .... 620 mi. .... 1,240 mi.

Land being fought over

Black Sea
Caspian Sea
Turkey
Mediterranean Sea
Iraq
Afghanistan
Iran
Pakistan
Saudi Arabia
Arabian Sea
India
Bay of Bengal
Indian Ocean

Map key

1 Armenia
2 Azerbaijan
3 Bahrain
4 Bangladesh
5 Bhutan
6 Cyprus
7 Georgia
8 Israel
9 Jordan
10 Kuwait
11 Lebanon
12 Maldives
13 Myanmar (Burma)
14 Nepal
15 Oman
16 Qatar
17 Sri Lanka
18 Syria
19 United Arab Emirates
20 Yemen

Israel

Israel is on the Mediterranean coast. The modern country of Israel was founded in 1948.

Jerusalem

Key facts

Size: 80,220 sq. mi.
Population: About 5 million
Currency: Shekel
Main languages: Hebrew, Arabic

Capital city: Jerusalem. This ancient holy city is about 4,000 years old. It is a center of Judaism, Christianity and Islam. There are many important historic sites from the Bible and the Islamic holy book, the Koran.

Jerusalem

Landscape: The hills of Galilea are in the north of Israel, and there is desert in the south. In between, there are fertile plains. The longest river is the Jordan (659 mi.), which flows into the Dead Sea. This sea is so salty that if you swim in it you cannot sink. Mount Meiron (3,963 ft.) is the highest point.

The Dead Sea

Places to visit: The history of this area goes back to about 2,000 B.C. There are many religious sites including the Wailing Wall, the Church of the Holy Sepulchre, the Mount of Olives and the Dome of the Rock. The town of Beersheba stands on the spot where Abraham is supposed to have pitched his tent 3,800 years ago.

Industries

Engineering
Electronics
Chemicals
Textiles
Fruit

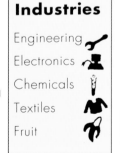

The Tomb of Absalom on the Mount of Olives

Saudi Arabia

Saudi Arabia covers about four-fifths of the Arabian peninsula. It is more than four times as big as France.

Riyadh

Key facts

Size: 829,995 sq. mi.
Population: Over 14 million
Currency: Saudi Riyal
Main language: Arabic

Capital city: Riyadh. This modern city is built on an ancient site. It is the home of the ruling Royal Family.

The Bedouin people wander the desert in search of grazing for their animals

Landscape:
Much of the country is barren desert scattered with oases. The Rub Al-Khali ("Empty Quarter"), in the south, is the largest sand desert in the world. The highest point is in the Asir Range (10,279 ft.). There are no rivers in Saudi Arabia!

Giant sand dunes in the "Empty Quarter"

Places to visit: Only Muslims are allowed to visit the sacred cities of Mecca and Medina. The Prophet Muhammad was born in Mecca and Muslims pray toward it, wherever they are in the world. Other ancient sites include camel markets, potteries and salt mines dating back 5,000 years.

Industries

Oil
Natural gas
Farming

The sacred city of Mecca

Iran

Iran is in Western Asia, east of the Mediterranean Sea. It borders the Caspian Sea.

Tehran

Key facts

Size: 636,293 sq. mi.
Population: About 57 million
Currency: Rial
Main language: Persian (Farsi)

Capital city: Tehran. The old city gates are in the old part of Tehran. It has one of the world's largest bazaars, where you can buy everything from carpets to silver and exotic spices.

A holy mosque

Landscape: Most of the cities are near the Caspian Sea, where the land is fertile. The rest of Iran is barren desert where much of Iran's oil is found. There are mountains in the western part of the country. The highest point is Mount Demavend (18,387 ft.).

Oil wells in the desert

Industries

Oil
Farming
Crafts

Places to visit: Iran was once called Persia. It was on the Silk Route, an important trail for merchants bringing silk and spices from the east. There are many ancient cities, museums and remains of the Roman and Persian Empires.

The Bakhtiari people spend the summer in the Zagros Mountains and the winter in the lowland areas of Iran

India

India is in Southern Asia between the Arabian Sea and the Bay of Bengal.

New Delhi

Key facts

Size: 1,269,338 sq. mi.
Population: Over 870 million
Currency: Rupee
Main languages: Hindi, English

Capital city: New Delhi. This modern city stands beside Old Delhi, where there are many ancient streets, temples, mosques and bazaars. The biggest city in India is Calcutta. Nine million people live there.

Old Delhi

India is the seventh largest country in the world, with the second biggest population after China.

Landscape: India is separated from the rest of Asia by the Himalayan mountain range in the north. In the north, many people live on the huge river plains of the Ganges and the Brahmaputra. Kanchenjunga (28,210 ft.) is the world's second highest mountain. The Brahmaputra (1,802 mi.) is the longest river.

Ganges

Industries

Farming
Chemicals
Electronics
Oil

Places to visit: There are many temples and palaces from the days when maharajahs ruled India. The most famous monument is the Taj Mahal.

The Taj Mahal

Ni hao!
Cantonese

Haere-mai!
Maori

This vast area includes China and its neighboring countries. Over one-fifth of the world's population lives in China. Asian Russia (to the east of the Ural Mountains) is in this region.

Malaysia and Indonesia (a group of thousands of islands) stretch down towards Australasia. This region includes Australia, the world's smallest continent, and New Zealand.

Chinese junks in the Pacific

Key facts about Northern & Eastern Asia

Size: About 12,399,601 sq. mi.
Largest country: Russia (about 4,835,516 sq. mi. in Asia)
Longest river: Chang (Yangtze) (3,915 mi.)
Highest mountain: Mount Everest (29,030 ft.)

Key facts about Australasia

Size: About 3,243,240 sq. mi.
Largest country: Australia (2,967,892 sq. mi.)
Longest river: Murray (1,609 mi.)
Highest mountain: Mount Wilhelm (14,794 ft.)

Landscape

Asian Russia has vast plains, with mountains in the far east and south. There is tundra in the Himalayas, with grassland and desert in central China.

Tropical rainforest stretches from southern China down through Malaysia, Indonesia and New Guinea. Some of this forest is still unexplored.

An erupting volcano

Part of this area is in the "Pacific Ring of Fire." It is called this because it has so many active volcanoes.

Much of Australia is dry grassland, with few trees. Australians call this the "outback" or "bush." There are also low mountain ranges, rainforests and deserts.

Weather

In the Himalayan mountains it is cold all year round. Farther south the weather is warm all the time. The monsoon season brings heavy rainfall to South-East Asia. Hurricanes (also called typhoons) often blow in this area. They are giant, spinning masses of wind and rain, sometimes stretching as wide as 300mi. across.

Hurricanes can blow at over 90 mph.

Malaysian — Selamat pagi!

Japanese — Ohayo gozaimasu!

Australian — G'day!

Economy

Eastern Asia is mainly farmland. Much of the world's rice is grown here. Rubber is also exported from here, all around the world. The rubber is made from milky sap, called latex, drained from trees.

Australia and New Zealand are important sheep-farming centers. In both countries, there are far more sheep than people. The sheep's wool and meat are exported to countries around the world.

Japan is one of the most powerful industrial countries in the world. It is famous for its electronic goods, such as computers and music systems. Japanese cars are exported all around the world.

Main industries

Farming
Mining
Oil
Shipbuilding
Forestry
Sheep

Map key

1 Brunei
2 Cambodia
3 Hong Kong
4 Kyrgyzstan
5 Laos
6 Macao
7 North Korea
8 Papua New Guinea
9 Singapore
10 South Korea
11 Taiwan
12 Tajikistan
13 Thailand
14 Turkmenistan
15 Uzbekistan
16 Vietnam

Map of Northern & Eastern Asia & Australasia

Japan

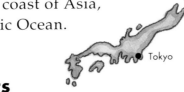

Japan is a chain of islands off the east coast of Asia, in the Pacific Ocean.

Tokyo

Key facts

Size: 145,833 sq. mi.
Population: About 125 million
Currency: Yen
Main language: Japanese

Capital city: Tokyo is one of the world's most crowded cities. The Imperial Palace, where the present Emperor still lives, is here.

The Imperial Palace

Japan has many islands. The four main ones are called Kyushu, Shikoku, Honshu and Hokkaido.

Landscape:

Two-thirds of Japan is mountainous. The only flat land is along the coast. There are over 200 volcanoes, half of them active. The highest point in Japan is Mount Fuji (12,389 ft.). The Shinano-gawa (228 mi.) is the longest river.

Mount Fuji

Industries

Vehicles
Electronics
Farming
Shipbuilding

Places to visit: There are many temples dedicated to the god Buddha, historic sites and ancient palaces. If you visit Japan you are likely to see traditional crafts, and rituals such as the tea ceremony.

A Japanese temple

China

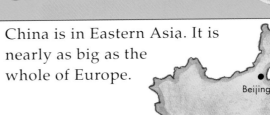

China is in Eastern Asia. It is nearly as big as the whole of Europe.

Beijing

Key facts

Size: 3,705,386 sq. mi.
Population: Over 1,150 million (about one-fifth of the world's people)
Currency: Yuan
Main language: Mandarin Chinese

Capital city: Beijing. There are beautiful palaces and gardens to see. The Forbidden City is in the middle of Beijing. This was once the home of the Chinese emperors.

The Forbidden City

Landscape: There is a high plateau in the west, with flat lands in the east and several very long rivers. Mountains take up about one-third of the country. Mount Everest (29,030 ft.), on the border of China and Nepal, is the world's highest mountain. The Chang (3,436 mi.) is the longest river.

Mount Everest

Places to visit: The Great Wall of China runs from east to west for about 3,977 mi. It is the only structure on Earth visible from the Moon. The ancient tomb of emperor Quin Shihuangdi is at Xi'an. It was guarded by thousands of life-sized clay warriors, called the "terracotta army."

Industries

Farming
Minerals
Chemicals

The Great Wall of China

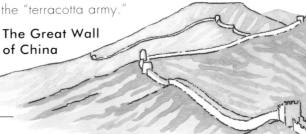

New Zealand

New Zealand is in the Pacific. It is made up of two main islands, North Island and South Island.

Key facts

Size: 103,736 sq. mi.
Population: Over 3 million
Currency: New Zealand dollar
Main languages: English, Maori

Capital City: Wellington. Only about 120,000 people live here and around the magnificent harbour. Over 1 million people live in Auckland, at the northern end of North Island.

Landscape: Much of the land is mountainous, with rivers and plains. In parts of North Island there are volcanoes, hot water geysers and pools of boiling mud. The highest point is Mount Cook (12,350 ft.). The Waikato (264 mi.) is the longest river.

Industries

Minerals	
Farming	
Wool	

A geyser spouting hot water

Places to visit: There is lots of spectacular scenery, including rainforests and glaciers. The Maoris were the first settlers in this country. Rotorua is the center of Maori culture. Many kinds of animals and birds, such as the kiwi, are found only in New Zealand.

Maori dancers

Australia

Australia lies 935 mi. north-west of New Zealand. It is the world's largest island.

Key facts

Size: 2,967,892 sq. mi.
Population: Over 17 million
Currency: Australian dollar
Main language: English

Capital city: Canberra. Most of the population lives along the eastern and south-eastern coastline. Australia is made up of six states and two territories, each with its own capital.

Sydney Opera House

Landscape: In the middle of Australia, there are several large deserts and dry grasslands. The Great Barrier Reef stretches 1,250 mi. along the north-east coast. The highest point in Australia is Mount Kosciusko (7,317 ft.). The Murray (1,609 mi.) is the longest river.

Ayers Rock, a famous landmark near Alice Springs

Industries

Farming	
Minerals	
Iron	
Engineering	

Places to visit: The original settlers in Australia were the Aborigines. Some of their cave paintings and historic sites date back to prehistoric times. There are wildlife parks where you can see koalas, kangaroos and other animals found only in Australia. The beaches are world-famous.

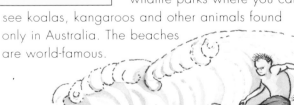

Index